How to Make Big Money in the Candy Vending Machine Business!

How to Make Big Money in the Candy Vending Machine Business!

Julian Aboulafia

VANTAGE PRESS
New York

FIRST EDITION

Published by Vantage Press, Inc.
516 West 34th Street, New York, New York 10001

Manufactured in the United States of America
ISBN: 0-533-09651-0

Library of Congress Catalog Card No.: 91-90981

1 2 3 4 5 6 7 8 9 0

To my family

Contents

How to Make Big Money in the Candy Vending Machine Business!

One

General Overview

Gumball vending, also known as bulk vending, is a great way to create a second income. It is an opportunity that you can create as quickly or as slowly as you wish. It is possible to create an extra $100 per month or as much as $10,000 per month. This all depends on how much time and effort you are willing to put toward this fantastic opportunity.

Let me say from the start that all of the information I give you contains extremely conservative numbers. I am not selling any "blue sky" or false hope. I am only giving you information as it has worked, and is working, for me. I am a man who has had to start all over in life at the age of thirty-eight. I am glad that I did so much "homework" on the vending business because I might have overlooked this great opportunity. I am equally glad that you are also willing to at least look into this idea before making a decision about it. You will find that this particular business can be controlled right from the start to be very large or very small and still earn you that "other income" you are looking for.

What Is It?

Bulk vending is that little gumball machine that you probably see at the entrance to different retail stores in your neighborhood. Whether in a grocery store, pizza store, dry cleaners, video-rental store, or auto repair or muffler shop, if you look closely, you will spot these machines all around town. Some of the machines have only gumballs in them. Some of those machines have three or four duplicate segments that carry a variety of candies like M&Ms, Skittles, Runts, Boston Baked Beans (BBB), and mints. Some machines even offer small toys, balloons, and other novelty products.

These candies are very popular, and the companies that produce them spend millions of dollars for TV and newspaper advertising and tie-in promotions for them. To show you the collective power of one of these companies, M&M/Mars Incorporated, a company out of Hackettstown, New Jersey, recorded $9.2 billion in sales in 1988. Within the entire vending business in the United States, chocolate and candy sales represents 62 percent of the industry. Recently, for example, I saw a promotional offer for a free video movie for purchasing four bags of M&Ms and a pizza from a national franchise-type store. This type of advertising works wonders for you!

As you read through the information in this book, you will get in on the secrets that the big-time vending "kings" use to make large sums of money (in cash) each month. If you only have weekends free and are willing to utilize this time, you can bring in as much as $300 to $500 extra per month with an investment of as little as $500 to $1,500. If you do not have this much money to invest, I will show you how to actually start your vending business for much less. And I *will* show you how.

Vending Machines? Are They Really Used So Much?

Yes! Most of us are not really aware of how many machines there are unless we start to take notice of them. There are many different types of vending machines, and their usage is as varied as our own needs.

To mention just a few different types, there are snack machines that vend candy bars, small bags of chips (like potato or corn chips), cookies, rolls of breath mints, packs of chewing gum, and other types of snacks that need no refrigeration if indoors. Then there are the cold snack vending machines that vend items like sandwiches, fruit, yogurt, and similar items, which do require refrigeration. Then there are cigarette machines, postage stamp machines, soda vending machines (both in the cup or in the can), newspaper machines, and even the ATM (Automatic Teller Machine) at the banks and convenience stores (like 7-11). The ATM machines are a form of vending machine that doesn't take coins but, instead, takes plastic custom-coded information cards that can take currency or checks and give you currency or transfer your holdings from one account to another.

Some of the vending industry's newest types of vending machines I have recently spotted are coin-operated FAX vending machines and even aluminum can recycling vending machines, which pay you in cash for the cans you deposit with the machine. If these types of vending machines are not in your area yet, they soon will be. As our society becomes more "high tech," more automated vending machines are showing up. The secret behind vending machines is: **"Many machines doing a little amount of business each."** This concept is exactly what Coke and Pepsi are using. Coke is selling 17 million gallons of Coca-Cola each day around the world, which is an $8 billion per year business. This has become a business foundation in our great nation. Theoretically, it is the idea of creating as many outlets for your product as possible. I call this "the **PRINCIPLE OF DUPLICATION.**" Thus the "franchise theory" spread across this great land and now we can get Domino's Pizza from Florida to California, delivered right to our front doors "in thirty minutes or less." It is this same theory that has proven itself over and over again that we will be working with.

Two

Getting Started

There are some personal considerations that you need to look at before you jump in. Let's go briefly through these few areas.

Time

You will have to reassign some of your time to "vending machine time." To get a better idea about time allocation, you need to "take stock" of all other home, work, and family responsibilities. Nobody said this was going to be easy, just very profitable. Also for your consideration are the hours of operation that the local stores are open in your neighborhood and surrounding neighborhoods. If your town is one that has all their stores closed on Sundays, for example, then Sundays would be a time to put into other areas. However, be aware that large chain stores that have outlets in your town probably have Sunday hours. Restaurants are usually open seven days a week. Each person is an individual with his own time utilization. Each person will use time differently, and some people work faster or slower, getting more work, or less work done in the same amount of time. So, recognize that you will need to use your time very carefully in each step of this vending machine business. Time-allocation is a very individually personal issue, and I just want my readers to realize that you will need to devote the necessary amount of time to vending, in order to gain a good start in this business, as well as a reasonable ongoing schedule.

Money

The amount of investment that you will need to allocate will depend on four things:

1. How much money should you invest toward this business? Remember, the amount of money that you invest, yes, *invest,* will return **monthly income**. Because of the **monthly** return, this money should be high upon your list of priorities. There are more advantages to this business besides monthly income. There also are tremendous tax advantages within this business. Some vending machine manufacturers will extend you credit terms to purchase their products, which involves their approval of your credit application.

2. How many vending machine "segments" and how much candy/gumballs/novelty items can I buy for my money? A candy-machine segment refers to one complete machine. This includes a globe where candy is kept (top), a coin acceptor (middle), and a place where coins are stored (bottom). It does not include the stand that the machine rests on. Most vending machines are sold in individual segments, and are placed two, three, four, or more segments on one stand. Some other machines have three segments encased in one plastic cover. These machines can cost from $45 each to as much as $350 each. The price of the machines depends most on whether the machines are new or used when you purchase them.

In considering the new machines, remember they are all clean and well-painted and need minimal assembly to be ready to locate in the business. Usually, just filling the globe with candy and attaching the machine to the stand is all you will have to do to prepare that machine for vending.

Buying used machines can save you a lot of money, but can cost you in time. Used machines will require cleaning, reconditioning or painting as is necessary. Usually, you will have to order parts for the machines from the manufacturer. The parts usually do not cost very much and are readily available for the ordering. All the above things need to be considered when you decide between new or used machines.

My recommendation is to spend approximately $50 for each machine, *including* its stand.

3. Buying bulk vending machines. The single most comprehensive publication in regards to specifically bulk vending machines and for the vending industry in general is a magazine called *Vending Times*. Write to: VENDING TIMES, INC., 545 Eighth Avenue, New York, New York 10018. Subscriptions are $30 per year (in the United States) or $3 for a sample issue. I highly recommend that you *immediately* send your check to them for a subscription. I have found *Vending Times* to be a great asset.

Purchasing new machines. You will find the larger, more reputable manufacturers of bulk vending machines advertising in the *Vending Times*. These are industrial machines made to be used for many years and still hold their resale value. To buy new bulk vending machines, I suggest the following reputable companies:

ASTRO-OPERATORS VENDING MACHINE SUPPLY COMPANY
7242 SCOUT AVENUE (UNIT A)
BELL GARDENS, CALIFORNIA 90210-4915
(213) 928-2538

ADVANCE MANUFACTURING COMPANY
11760 ROSCOE BOULEVARD
SUN VALLEY, CALIFORNIA 91352
(818) 767-5466

NORTHWESTERN SALES AND SERVICE CORPORATION
446 WEST 36th STREET
NEW YORK, NEW YORK 10018
(212) 564-6467

A & A COMPANY
PARKWAY MACHINE CORPORATION
1930 GREENSPRING DRIVE
TIMONIUM, MARYLAND 21093
(800) 638-6000

The following things must be taken into consideration:

I suggest that you try to buy your machines from a local source or within one-day, round-trip driving distance. This will avoid shipping charges on your machine order.

The quantity of machines you order per order directly affects the price of each machine. For example, ordering one to ten machines might cost about $47 to $50 per machine. However, if you can order ten to twenty-five machines at one time, your cost per machine will drop to approximately $40 to $45 per machine. Now, if you are in the position to order twenty-six to one hundred machines, then you can probably get each machine for $35 to $40 each. So, your advance planning is the key. **Buy as many as you can at one time to reduce your costs.**

For **used** bulk vending machines, I suggest you check local newspaper classified ads under "machines, vending." Sometimes the categories "miscellaneous," "Coin-Op/Vending Machines," or a similarly named category is used. Not only the larger newspapers, but also the small trading newspapers, like the *Recycler* or *PennySaver* or *Nifty Nickel* or some similar publication will have advertisements that individuals will run in order to liquidate some or all of their bulk vending machines. Then you might really get lucky and find someone who just wants you to take their machines off of their hands and will sell them to you for a fraction of what they are really worth! Just be aware that the used machines that you buy might need extra cleaning, parts, or painting to get them ready for you to use.

It is also important to remember that the cost of bulk vending machines does not include the cost of the machine stands. These stands can run from $15 to $40 each, and can accommodate from one to four machines for each stand. So, when you are doing an approximation on your start-up costs, don't forget to include the cost of your stands. You will have to decide how many machines you will want on each stand before you order the stands. I suggest three machines on 95 percent of your stands if you will be using the pole-and-base type of stands. If you use the TV tray-style stands with the chrome-plated legs, figure four machines on each set-up stand, for 95 percent of your machine order.

Before you start making checks or money orders out to the vending machine company you choose, I suggest you call three different companies and ask to speak to someone in the machine sales department. Don't be afraid to ask for prices of each machine; the difference of prices for ordering ten machines, twenty-five machines, fifty machines, one hundred machines, and compare the difference in cost of each machine when purchased in different quantities; different sizes of their machines; photographs of what the machines look like; shipping costs of each machine to your address, and also the availability of parts, price lists of spare parts, as well as costs on candy per pound, and price of shipment costs of candy to your address.

Also, consider if you have enough space around the house to handle the room that your purchase of the machines, stands, and candy will take up. If not, look into an inexpensive store & lock space you can rent by the month to hold all the vending supplies you will be purchasing.

Will you need to borrow or rent some type of small truck or van to transport the machines, stands, and candy from the place of purchase to the place you have designated to keep all these supplies? If so, make some phone calls, and figure in these costs in advance.

The answers to all of these questions are very important because they give you the ability to more accurately predict your start up costs for buying the machines, stands, and candy to fill the machines with. This is the type of cost projections that you would need, in order to be confident when writing those checks for your first purchase.

The candy supplies do not have to be ordered from the company that you order your machines and stands from. There are many companies that carry the same type of candies, and their prices are all pretty much standard. Bulk vending machine manufacturers supply candy to the machine buyers strictly out of convenience to the buyers. This entices machine buyers to come back to the same manufacturer and gives the manufacturer a better chance to sell you, the buyer, more machines. Check in your phone book under "Vending Machine Supplies," or even "Candy," to find suppliers of BULK VENDING CANDY . . . the candy that is packaged twenty, twenty-five or thirty, even thirty-five pounds per box. The candy is not wrapped in pounds, but in five-pound packages or more. The exception is M&M plain or M&M peanut candies and Skittles, which are packaged in one- or two-pound bags. I recommend the one- to three-pound bags for keeping the chocolate fresh. Another item to remember is that all candy has a certain shelf life, which is usually a date stamped on the side of each box as it leaves the candy company. Be aware of these dates. Order only the amount of candy you need to fill up the machines. Try to avoid stockpiling any candy, so that it does not get stale.

To get an estimate of how much candy you need to order, remember that the plastic globes that hold the candy should not be totally filled up. The clear plastic globes should be, at most, only three-quarters full at any time. This promotes sales more readily. If the globes are completely filled with candy, the customer's psychological tendency is to wonder, *Why isn't this candy selling? Is there something wrong with it?* That can cost you sales! So figure how many globes you will be filling with each different type of candy, multiply those two numbers together, and this will give you the approximate amount of candy you will need to order. Usually, just for estimation, you can figure if each globe will hold ten pounds of candy, you should fill only to the five- or six-pound level. The only reason that you will need to fill to the top of the plastic globe is because you have an extremely "HOT" location, whose sales prove to be very high, and every time you get to this location to service these particular machines, the machines are empty. This is a clear indicator that this particular location requires your filling the plastic globe all the way up. Remember, the smaller amount of candy you have in the globes, the better the candy appears to be selling to the customer, which will entice more sales. Also, the lower the level of candy is, the more often will fresh candy be in the machine. Usually, the fresher the candy inside the machine is, the higher amount of sales the fresh candy promotes, especially to the employees.

Bulk vending machine stands. The machines' stand usually consists of a circular base that is heavy enough to weigh the machines down and prevent them from becoming top-heavy so they will not tip over. The stands also hold the machines a convenient height off of the floor (about three feet) for maximum accessibility. Attached to the round base is a pipe or pole. Either a steel pipe or a hollow aluminum pole that had been painted. Using a cross brace, the machines attach to the top of the pipe or pole.

Another type of stand, newly available, is made of chrome-plated, aluminum tube material and resembles two upside down "U"-shaped pipes that spread out like the four legs of a table, similar to a TV food tray. You may have noticed these types of stands in supermarkets with six, eight, or ten machines on them. Usually, very large machines are placed in supermarkets and are larger than the machines you will be working with.

There are two other types of stands available. One is a wall-mount stand, which must be mounted by drilling holes in the wall where it is to be located. This makes it impossible to easily relocate the machines; more important are the legal implications of attaching your machines permanently to someone's property. The second type is the counter-top stand, which will put your segments on a counter, which in most businesses is one of the most valuable spaces that the business has to display their high-impulse items. Using this space for vending machines is usually not allowed. I discourage both of these types of stands.

4. How much money can I make on bulk vending machines? Here is a chart supplied by one vending machine manufacturer. This will answer the question about the amount of profit available through bulk vending machines. Remember that candy prices change, but this chart was published at the start of 1990. (The following information is based upon twenty-five-cent coin vending machines.)

CANDY NAME	PRICE PER/LB.	INCOME PER/LB.	ESTIMATED PROFIT/LB.	CANDY COST
M&M plain	$1.70	$6.00	**$4.30**	29%
M&M peanut	1.70	4.13	**2.43**	42%
Reeces Pieces	1.89	6.00	**4.11**	32%
Skittles	1.69	6.00	**4.31**	28%
BBB	1.27	4.75	**3.48**	27%
Runts	1.15	4.75	**3.60**	24%
Mike & Ike	.99	4.50	**3.51**	22%
Hot Tamales	.99	4.50	**3.51**	22%
Chickle gum	1.15	4.75	**3.60**	24%

This is a small sampling of the over forty different types of bulk candies available for you to use. Remember that the amount of money that each pound of candy will sell for, directly depends on how many pieces of candy you give for each quarter. This does not even touch upon the novelties and toys. (Check with your specific machine manufacturer for recommended candy settings on each type of candy, and they will tell you how many pieces of candy you should give for a twenty-five cent/ten cent vending sale.)

Labels

There are two types you will need. The first type identifies you as the owner of the machine and consists of self-adhesive labels. These labels cost about $10 for 250 of them. They are printed with your last name, or your business name, and your business address and, most importantly, your *phone number.* This is perhaps the only way that a store owner/manager can get in touch with you in case the machine should malfunction or be sold out. Or perhaps the business is closing or moving and your machines have to be removed. I am sure you would rather get your machines from the store while its doors are open for business than try to contact the shopping center owner or management company and wait weeks for the area representative from the management company to send a man to open the door for you to take your machines out. All this time your machines could have been in some other location selling candy instead of sitting in the store of a closed and locked-up business, with you having no access to them. These labels can give the owner/manager the only link to you and will be the only communication link that anyone can have to get in touch with you about your machines. I suggest the self-adhesive labels be stuck on the back or the side of the bottom portion (coin box) of your machines.

The other type of labels you should have consists of candy labels that are placed in the front of each globe; this shows the customer what candy is inside the globe. You can best profit from the millions of dollars that the candy manufacturers spend in logo and trademark recognition for consumers to identify their candy, thereby encouraging more sales. These logo signs are available from the same place you buy your BULK candy. Often the salesman will give these signs to you free, or for five to twenty-five cents each.

Three

The Three Ls of Vending

The next area that we need to become acquainted with is finding places where your vending machines can be put so that they can do some serious selling.

Locating

This is one of the very best aspects to owning vending machines. They sell while you are somewhere else doing something else, and you need not be present for the sale to take place. If you understand the PRINCIPLE OF DUPLICATION, you will see just how exciting this can be for you. Imagine you are eating dinner at home or are even at work on the good old nine-to-five, and while you work, people are putting coins into YOUR vending machines all over town, without you having to be there, and those coins really do add up! I guarantee that this will become as exciting for you as it has been for me.

Locating is: the action of finding a business (either product- or service-oriented) that will allow you to install, service, and maintain vending machines on a regular timely basis within the business's physical space. The first rule of placing bulk vending machines into a store is "LOCATION! LOCATION! LOCATION!!!" The quality of the location determines the volume of sales from this specific vending machine. There are two general rules that you should keep in mind when you seek out locations:

1. The larger the established business is in terms of number of employees, the more sales you will make.
2. The greater the business's walk-in customer flow is, the greater the vending sales will be.

Don't forget that not only will walk-in customers buy products from your vending machines, but also the employees will be your best regular customers. (My first inclination was to immediately solicit every large retail store and service business in my area by mail.) Instead of doing this, I immediately went out and spoke with managers of the large successful chain stores around my neighborhood. In order to have some fun, make locating a "fun game." Go to locate when you are in the best mood. *Nearly*, I repeat, *nearly* all of these large retail food chains, drug stores, fried chicken, and ice-cream stores already have specific rules about these vending machines. The machines are either not allowed according to the by-laws of the franchise headquarters or else the arrangements have already been made between the large national bulk vending companies and the chain stores that *will* allow vending machines in their stores. Also, these large national bulk vendors can afford to split their sales with the chain stores on a fifty-fifty basis, but you cannot. Now that you know what *not* to do, the good news is that *not all* of the chain stores are manager-operated stores. Some of the franchise owners do indeed have the decision-making ability to allow your vending machines in their store. Do not be discouraged because there are literally thousands of stores and service-connected businesses remaining. And then again, there is the next town right near the one you live in, and then the next, and the next . . . get the point? Don't get "worked up" over ONE location so badly that if you don't get it, it hurts. Just forget it, and keep right on going.

Locations

Locations are: the specific businesses where you have placed bulk vending machines to sell product to the customers and employees of that business. All the bulk vending machine owners or operators that I have talked to seem to have their own ideas about which types of businesses are best for bulk vending machines to be located in. I have compiled the following list from a combination of all their answers. The type of businesses closer to the top of the list have generally been found to produce the highest amount of sales, whereas the businesses closer to the bottom of the list have been found to produce lower amounts of sales. All the businesses listed here are doing an acceptable amount of volume.

I need to caution you in this area of preferred locations that from one town to the next, or from one state to another, no single type of store will be the "best" type of store to locate your machines in. This business is a numbers game; hence, the more machines you have out there, the more sales income you will receive. You should expect sales to vary from one location to the next in all geographic areas. The public's preference for different types of candy and novelties will vary also.

Recommended Vending Location List

Pizza Parlors
Pizza Hut
Round Table Pizza
Shakey's Pizza
Straw Hat Pizza
Any Pizza Parlor

Video Stores
Tape Rental Store
Video Depot
Warehouse Rental
Blockbuster Video
Any Video Store

Fast Food Stores
Burger Palace
Tastee Freeze
H. Salt Fish
Hot Dog Stands
Any Fast Food

Restaurants
Love's B.B.Q.
House of Pancakes
Howard Johnsons'
Coffee Shops

Malls/Shopping Centers
Indoor Malls
Outdoor Malls
Farmer's Markets
Flea Markets
Swap Meets

Ethnic Stores
Any Mexican Store or Restaurant

Any Chinese Store or Restaurant
Any Ethnic Store or Restaurant

Commercial/Industrial Plants
Snack Shops
Employees' Cafeteria/Lounges
Near food-vending machines
New & Used Car Dealers

Miscellaneous
Gas Stations
Drug Stores
Clothing Stores
Dry Cleaning Stores
Auto Parts Stores
Delicatessens
Bakeries
Laundromats
Banks
Sandwich Shops
Post Office Box Stores
Auto Dealerships
Auto Mechanics Shops
Muffler and Auto Specialty Shops
Record/Cassette CD Stores
Hair-Styling Shops
Hardware/Lumber Stores
Large Real Estate Offices

And into infinity we can go from here.

(NOTE: Some franchises are individually owned; some are company owned. DO NOT ASSUME ANYTHING! **ASK!**)

Locators

The final "L" is for **locators**. Let me preface this next part by saying that not everybody in the bulk vending machine business needs to use the services of locators. I was involved in other businesses and really could not afford to randomly take days off and walk around my neighborhood to speak to all the local business owners to place vending machines. When I did have the time, I wasn't really having very much success at the art of locating. It was at this time that I realized that I need help. The plan I used was to start locating machines in businesses around my home, both in the closest businesses and within the closest shopping centers. This built my foundation, or base of operation, close to me for easy accessibility.

The use of locators is optional. However, the way they work and the amount of money they charge is what you should know about. **Locators** are: Salespeople who will scout the areas you designate and speak with as many business owners/managers as possible to secure as many vending machine locations as possible in one day. The locator's commissions are on a "per location" basis. The fee for bulk vending machines usually varies from twenty-five dollars per location to about fifty dollars per location. The most I have paid for locations in Southern California was thirty dollars. Locators can make as much as two hundred dollars to find a single location for a private pay phone in a shopping center. They also locate soft-drink machines, snack machines, and other types of vending machines. The smallest fees are for the smallest vending machines, which are bulk vending machines. There are two major categories to this service. In-town locators and out-of-town locators.

In-town locators. In-town locators are the least expensive to use. These are people who live in your same town and use their car to drive from shopping center to shopping center (or business district to business district) and go from store to store within the center to find suitable locations for your machines. It is customary, but not required, to offer the business owner/manager a small percentage of the sales of the machine to be paid to the business each month or each service cycle. Vending machine owners I know of, or have heard of, offer to pay ten percent to fifty percent of the sales of the machines on a monthly or service-cycle basis. Personally, I have found that twenty percent commissions are very acceptable. In a few of my locations, I pay twenty-five percent. I really, as you should, try to minimize the higher commission percentages. There are many businesses that will let you put your machines in for no percent at all.

It is very likely to have different commission rates for different stores on your vending routes. When you have counted up the coins that were in the machines from this service visit, figure their percentage out and write the amount of commission in a two-copy receipt book, which can be purchased at any stationery store. Be sure to include the date and the name of the business on the receipt, and ask the manager or the employee entrusted to run the cash register to sign the book. Then, simply pay that person the proper amount of commission right on the spot and give him/her the top copy. Keep the bottom copy in the book for your tax records. The tax category "commissions paid" is a tax-deductible category, and you must keep a record of these payments. Be certain that you and your locator have a clear understanding of what commission percentage you are willing to pay the store owners/managers before the locator starts. Any confusion could cause serious problems.

I recommend that you or your locators carry a sample book made up of eight-inch by ten-inch color photos of your machines, with the globes filled to the top with candy and with candy labels shown. This can easily be done with a 35-millimeter camera and color film. These photos will do wonders toward getting locations and will help them decide what style of machine they want, if you have more than one style.

Another issue that you must establish with your locator is a make-good policy. A "make-good" is the promise of your locator to replace or "make-good" any location that he/she has gotten for you that did not finalize, for some reason. This sometimes happens, for a variety of reasons.

For example, the locator may have gotten permission from a store employee who claimed to be the manager. The locator had no reason to doubt the employee when permission was granted, the locator added this store to his list of stores he had accumulated that day, and you paid him for finding this location. When you subsequently go into the new location with your machines in one hand and your location agreement (covered in a later chapter) in the other, the real manager politely, but firmly, refuses to allow the installation. A reputable locator will realize that this sometimes happens to him and gladly do a "make-good" out of the next batch of locations he gets for you. This merely means that one of those new locations will be at no charge.

If a locator is really experienced, he will know which stores to avoid and which stores to work hard to get. This type of locator is worth every penny you give to him for his locating services. When the locator is giving you the list of the locations that he got for the day, be sure he gives you the following information:

1. Name of business.
2. Address of business.
3. How many machines to be placed at that location.
4. Specific candies (if requested by owner/manager).
5. Name of owner or manager who gave permission.
6. Note the date you bought the location from the locator.

You must keep a record of the locations that this locator has obtained for you. Put all of his locations in a folder or notebook of some kind so that you will not lose track of your make-goods from ths locator. You may have as many as two or three locators working simultaneously when your business gets much larger, so you must keep accurate records of all locations and all make-goods.

Out-of-town locators. All of the above rules that I have discussed apply equally to the out-of-town locators. The out-of-town locators must travel to your area before doing the same work. The cost, however, is quite different. If you employ an out-of-town locator, you are required to pay his transportation costs, food expenses, hotel or motel expenses, and location fees. Needless to say, this can run up the cost per location; however, some of these locators are so successful at their work that you might want to use them. I would suggest that you consider out-of-town locators working for you only after you have achieved a certain level of success in the bulk vending business. I also want to mention at this point that the expenses of transportation, food, and hotel rooms are a negotiable item. Some locators will be willing to split the costs fifty-fifty. In any event, it is best to rely on in-town locators during your initial start-up stages.

Whether in-town or out-of-town, all locators have to make a decision about the possibility of putting your machine into a place where there may already be some bulk vending machines. This is referred to as **PIGGY-BACKING**. You have to be sure to tell your locator if you do or do not want your machines located in a business where other machines already are. Surprisingly, I have found the majority of these "piggy-backed" or "shared locations" to be very successful. There are a few things that need to be considered:

What is the current number of bulk machines at the location before I add mine? If there is only one other set of one, two, or three machines in the location, I will go ahead and put mine in because both the employees and the customers are aware of the other machines in this business, and I can also benefit by this awareness. If this location has more than one current machine set up, I will usually pass on this business location.

What types of candy or toys are in the other machines? You need to be sure that you can still put two, three, or four of your machines into this location and not duplicate any of the items that are in the other machines. Duplication means less sales for everyone.

Is there room enough for my additional machines? Can additional machines be put into this spot without over-encumbering the space? You need to visually inspect this situation to know the answer to this. Many times, your machines can be located in another space that is away from the other machines that are already in the store. Be careful, do not allow your machines to be put in a corner or in a space that has a reduced customer traffic flow. As a rule, if the machines cannot be seen easily by the majority of customers, the location will produce low sales.

Four

Hooking Up with a Charity

As you start becoming aware of the bulk vending machines around your town, take a little closer look at those machines. Walk over to them and notice that nearly every one of those gumball machines have some sign on them informing the reader that the use of this machine helps a charitable institution. There are several reasons that these bulk vending machines are usually associated with a charity.

1. Some charities will allow you to use their name on a small sign affixed to the machine in exchange for a monthly contribution made to the charity. Some charities actually have a vending machine division devoted to keeping track of all of the owners and their monthly contributions. Charities charge you a small fee each month for each machine. The fee varies from one dollar per machine per month to one dollar fifty cents per machine per month.

2. Your machines will be used more with that charitable organization's sign on them. This just happens to be a fact. People want to believe that the coins that are being deposited into these machines go directly to the charity named on the sign. Well, in fact, part of that coin does on a regular monthly basis. So these charity signs are very important for making sales.

3. Perhaps most important, if your machine happens to have a mechanical malfunction during its use, most people will just forget about it and walk away from the machine. The importance of this is that the customer who did not get the candy he/she was expecting will not make a big deal about it to the management of the store. One of the fastest ways to be asked to remove your machine from a location is for the machine to be a constant source of irritation to the employees working near the machine or in the same store and having to hear constant complaints about it.

It is a fantastic help that you can be to so many others just for that small monthly contribution. The charity hook-up actually serves as a singular purpose for wanting to place the machine in the owner/manager's store. Do you see what I mean? The sign itself becomes reason for the owner/manager to allow the machine in his store. And the contribution to the charity is really not so much money each month. Plus, of course, it really does help so many people who are dependent on that charity for various services. I strongly recommend you hook up with a charity in your town or state or even a national concern.

Some of these charities, to name just a few, are: The National Federation of the Blind; United Cerebral Palsy; Make A Wish Foundation; Kiwanis Club; Crippled Children's Society; Jerry's Kids, and many, many others. You can get all the phone numbers and addresses from the yellow pages under "CHARITIES" or "NON-PROFIT ORGANIZATIONS." In one hour or less, you can have three or four charities mailing information to you about their vending machine division. Look them all over, and decide which charity you would want to hook up with. If you already have a preference, good! Go with it!

Five

Keeping Track

So now you know where to look to buy bulk vending machines, where to get candy, what questions to ask of the candy and machine manufacturers, and you are armed with all the basic information you need to take the step of purchasing your first batch of machines. Whether you buy ten or one hundred, the information is the same. You have even been let in on how locators get their locations. Now, you make the purchase, find a truck to transport machines, stands, candy, and have a place to put them. Kick your locator into action to get you locations for these machines, and now you really need to keep track of all the places you will be putting machines.

There is nothing more frustrating than getting approved locations and then deleting a location because your locator, or you (if you are getting your own locations), have done everything right except make a list, or keep track of all the places that you have just placed your machines. You know it is out there, but you are really not sure just where.

As amusing as this might sound, I know a person who was getting out of the bulk vending machine business. I agreed to buy the machines, but he forgot all the different locations that the machines were in. One week later, I got a phone call from this same person to tell me about two locations that were not on the original list that I was given. The machine locations were inadvertently deleted from the list.

When you also get into a large volume of machines, you will understand how easy it is to forget that small location that has two, three, or even four of your machines. It is easy to lose track sometimes. A good "operators" policy to have is to do all the necessary paperwork before you place a location into operation. This will help you to prevent losing track of any machines.

There are a few forms that I believe will help you organize your business from the very start.

To start, there is the **LOCATION AGREEMENT** form. This very simple form gives you written permission to place your machines in another person's store. Let me say at this point that many machine owners do not even use any type of location agreement. They just use a handshake, and the machine gets put in. I believe that a simple form is the best assurance for any legal repercussions. Actually, this simple location agreement just puts into writing the fact that there has been permission given for your machines to be in another person's store. Here is a sample LOCATION AGREEMENT on the next page.

SAMPLE LOCATION AGREEMENT (8½″ x 11″ paper)

LOCATION AGREEMENT

WE, THE UNDERSIGNED, PERMIT OUR ESTABLISHMENT TO ALLOW VENDING MACHINES (BULK) TO BE PLACED ON THE PREMISES.

IT IS UNDERSTOOD THAT EQUIPMENT WILL BE MOVED WITH 30 DAYS NOTICE FROM EITHER PARTY. TITLE TO THE MACHINES SHALL REMAIN IN THE NAME OF THE VENDING MACHINE OWNER WITNESSED BELOW.

BUSINESS OR LOCATION NAME: ....
BUSINESS ADDRESS: ....
....
AUTHORIZED SIGNATURE: ....
(MANAGER/OWNER)
ACCEPTED BY: ....
VENDING MACHINE REPRESENTATIVE
FOR: ....
(CHARITY)
OWNER NAME: ....DATE: ....
ADDRESS: ....PHONE: (   ) ....
CITY: ....STATE: ....ZIP: ....

NOTES: ....

MACHINE NAME AND NUMBER OF SEGMENTS: ....
PRODUCT: ....
ADDITIONAL INFORMATION: ....

(not to scale)

You can see that this form is very basic. You should bring this form with you at the time the machines are being placed into the store and ask for the name of the person that was given to you by your locator when this location was secured. The owner/manager can read the form in thirty seconds and will sign. Then just keep all these forms in a three-ring notebook and photocopy more blank ones when needed. Feel free to modify this form; keep it simple.

The next form you will need to use is a type of **ROUTE CARD**.

The purpose of a route card is to keep information about all your locations in a sequential style. The route cards are small, approximately four inches by seven inches, and are easy to carry as you service your route. The cards have two holes on the top, bound by round rings, similar to the round rings used to hold your car and house keys. This binds the route cards together in a manner that will allow you to flip one card behind the last and therefore establish a route for you to follow in your servicing actions.

Each card has data about each complete machine location set up. Included data on the route card are: date of placement; machine type and amount of machines; name of business; address of business (including city, state, zip, and phone number). Also included are: name of contact (owner/manager); machine serial number(s); name of route this location is in; and finally columns of the date of collection and amount of money taken from machines. (In the upper right-hand corner of the card write the commission percentage the business gets.)

The **ROUTE CARD** looks like this:

MACHINES .... DATE PLACED .... % ....

BUSINESS ....

ADDRESS ....

CITY .... STATE .... ZIP ....

ROUTE NAME ....

CONTACT .... PHONE (.) ....

DATE/COLLECTED	DATE/COLLECTED
[1/15/90 / $67.25]	[/]
[/]	[/]
[/]	[/]
[/]	[/]
[/]	[/]
[/]	[/]
[/]	[/]
[/]	[/]
[/]	[/]
[/]	[/]
[/]	[/]
[/]	[/]
[/]	[/]
[/]	[/]

(not to scale) (Should be about 4 × 7 inches)

Remember, simplicity is the key. Get one of these typed up on a plain piece of paper and have them reproduced on heavy stock that will last for a couple of years, or even longer. Choose a copy store that you can solicit for a location! Offer them a 20 percent commission!

It is very important for you to understand that accurate analysis of the locations you are starting will rely upon keeping accurate records. I do not want this to sound very difficult, because it is not. By creating the last two forms that we talked about, the LOCATION AGREEMENT and the ROUTE CARD, you have actually backed up all information from one form to the other. Now, just in case you lose the route cards, or if for some reason the route cards get destroyed, the location agreements can be used to duplicate the route cards. All of the information on one form can be obtained from the other form. The only exception to this is the dates of the location servicing and the amount of the sales; this is recorded only on the route card and not on the location agreement.

In order to back this information up, I suggest you create a third form. This form is the **LOCATION GROSS INCOME RECORD**. Very simply, on a plain white piece of paper (8½ inches by 11 inches), record the same information that is on the route card. This serves two purposes: First, as a duplicate of the individual servicing dates and sales amounts from each location, and second, as a bookkeeping foundation for your end-of-the-year taxes.

Now, all you need, since you have just created your income books, is a list of your **BUSINESS EXPENSES**. These expenses include all machine purchases, machine stands, machine parts, candy, and even a certain amount of auto maintenance, gas, and the other tax advantages you should familiarize yourself with. I do not want to get into the tax field at this point, but most small businesses have great tax advantages. There is a wealth of information available specifically on this subject, and I recommend you buy a book or find one at the library.

As you can see, keeping track is not that difficult after all. With only four simple forms, you can document the foundation of your business. Incidentally, if you have a home computer, you will see advertisements in the *Vending Times* for not only accounting and tax programs, but specifically for vending programs tailor-made for your bulk vending business.

Six

Machine Maintenance during Route Servicing

It is important to establish a maintenance routine as you service your routes. Regular servicing means collecting your coins and is an exciting form of your payday. This is the payoff for all your work and patience. You can look forward to "payday" each month as you service your routes. I just want to take some time to share with you some of the successful actions that I have adopted into that servicing routine. Be aware that, for the most part, the time that you spend on your route to service your machines is usually the only time that maintenance can be done to your machines. This time spent with the machines is the best time to perform not only any maintenance that might be needed, but if you work smart, you will take another minute or two to perform some preventive maintenance. This will ultimately limit your service calls, and it will prolong the life of each machine and possibly the longevity of your business.

In order to give you some idea of a "servicing routine," I will have to bring into the picture a few tools that you will need to bring a professional approach and result to your efforts. The tools will vary according to the machines that you choose to buy. For this reason, you will see that I will recommend one thing for one type of machine and another remedy for other types of machines.

Thankfully, bulk vending machines are very basic in their operation, because I certainly am no mechanic and even *I* find them easy to work on. Don't get all bogged down in the workings of the machines because you do not have to be a "field technician" to assemble or fix these machines. You can always do a thirty-minute exploratory examination with one of your machines when you first buy them, to gain a basic knowledge of how they work. If you feel as though you really need to, you can get literature "exploded" diagrams and instruction pages or a booklet that is made available by the same manufacturer who makes the machines. I personally have never found any great need to have these books. The machines have only a few moving parts in the coin mechanism, and all other parts are stationary. The stationary parts just fit together like a puzzle. Just assembling the top part to the bottom part and then putting the machine onto its stand will teach nearly all you will have to know about the machines, with the exception of the coin mechanism. You should dissect one of these mechanisms, to familiarize yourself with the parts. Then you will see how very basic and simple the machines are.

Before you start your route servicing, you will be doing what we just referred to as a basic dissection on the coin mechanism. At this point I want to remind you that the manufacturer of your machines will be the best source to ask about specifically which settings your candy proportioner should be set at to vend each different type of candy or novelty. In most of the proportioners, the number or letter settings will vary, so it is very difficult for me to give this information to you accurately. What is important is that your candy should return three times what you paid for it when it is vended in proper amounts.

For example, there will be a different setting for a small candy like M&Ms (plain) and a different setting for M&M peanut. Likewise, the larger candies like Mike & Ike and Hot Tamales will have still different settings. So remember to get the manufacturer's recommended candy settings to vend proper amounts for each coin. Usually the manufacturers have already printed charts that will tell you just what setting to use for each type of candy. This chart is usually available at no cost or for twenty-five cents. Follow these settings exactly. This will govern the amount of profit you make on each pound of candy you buy and vend.

When you are ready to check up on those first few locations that have been placed for at least one month, here is what I recommend you bring for servicing the machines in your route:

1. A dozen small clean rags (approximately 12 inches × 12 inches).
2. A plastic half-gallon water container filled with clean water.
3. The keys to your machines.
4. The two-copy receipt book and pen.
5. Scrap paper to add up sums of counted coins.
6. Your route cards.
7. A can of spray lubricant (like WD-40).
8. Any screwdrivers or pliers that you used when assembling the machines.
9. Your stock of candy or novelties you will need to refill your machines.
10. Coin bags (Canvas. Ask for these at your bank.)
11. (Optional) Can of ant and roach spray, just in case you need to spray around the base of the stand. **(Never spray near the candy.)**

12. Plastic disposable gloves (without any powder on them).

You can carry these things in some type of plastic tray that has a handle, or a plastic bag, and take two clean rags. Wet one of the rags with the water you have in the half-gallon container. Put the water back into the trunk. Now you have one wet rag to wipe off the machines and one dry rag to dry the water off of the machines. Put the two rags into your plastic tray or plastic bag. Try to keep them separated so that the dry rag will not get wet.

Also, put into the tray or plastic bag the following:

1. The keys (unless you carry them in your pocket or in your hand).
2. The receipt book.
3. The route cards.
4. Some scrap paper and pen.
5. The can of spray lubricant.
6. One canvas coin bag.
7. The ant and roach spray, if needed.

Now you are armed with all your basic tools you will normally need to service your machines.

To start, when you get into the store, it is important that you act confident about what you are doing. Know that if there is something wrong, you can fix it. Inform the owner/manager or clerk in the store that you are here to "service the machines." This will explain what you are doing around the machines and also will open the conversation up for the clerk to tell you if there has been any problem with the machines, which he/she will do at this time. If there was a problem, just listen to what the person has to say. He/she will probably tell you exactly what is wrong by his/her description. You will come to realize that there are only about half a dozen things that can possibly go wrong with the machines anyway. Thank him/her for informing you of the problem (if any).

Now observe the machines closely for about five or ten seconds. Look at the candy levels to determine how much has sold. Look at the candy condition. Be sure that the heat-sensitive candy (such as M&Ms) is not cracking. If they are, the machines will have to be relocated to a cooler spot in the store or where there is not so much sunlight directly hitting the machines. (Remember that the heat-sensitive candies are in a type of greenhouse. The sunlight gets intensified inside the globe.)

If cracking is evident, just visually look around for a better location and ask the clerk if you can move the machines to the spot that you have picked out. Explain about the "greenhouse" problem with the candy and ask permission to move the machines. If it cannot be done, do not get discouraged; just change the candy to a less heat-sensitive candy. For example, from M&Ms or M&M peanut to Candy Coated Peanuts (Boston Baked Beans), or possibly Runts. If you are permitted to move the machines, make the move and thank the clerk for allowing you to do that.

Now that you have inspected the condition of the candy, look particularly at the complete outside surface of the machines. Try to spot if someone set a soda can on the top of one of the machines and the can left a round, sticky ring. Are there dozens of fingerprints on the plastic globe that are making your brightly colored candy look dull? Are either the stand or the tops of the machines dusty or full of dirt? You see, just a small amount of wiping down will correct many bad appearances. Now, you can wipe off the fingerprints, dust, soda marks, and dirt with the damp cloth rag, and dry the machines and stand with the dry cloth rag. To dry, especially the plastic globe, will prevent drying streaks and marks, and make your candy appear bright and appetizing. This will enhance your sales and keep your machines in better shape. All this wiping can be done in thirty seconds or less. Now that is not so bad, is it?

Now get our your keys, and proceed to empty, AND COUNT, the coins that are in each segment. As you proceed from one segment to another, write down the amounts on a scrap paper you can carry in your pocket, so that if someone interrupts you while you are counting, you will not forget the amount of coins you just counted. Then add all sums of coins up to their dollar value. Write this amount in two places: first in the proper space on the route card, along with today's date, then in your two-copy receipt book. Also in the receipt book, write the name of the business, the date, and the total sum that you are giving to the store owner or manager as "vending machine commissions" (usually 20 percent). For example: "March 27, 1990. Twin Valley Video. Vending machine commissions $5.25."

Now go over to the clerk or owner/manager and count out five dollars and twenty-five cents in coin and ask him/her to please sign the book. After the book is signed, tear off the top copy and give this copy to him/her. The bottom copy is for you to keep for your records.

Finally, the last thing you do is figure out how much of each type of candy you will need to bring the levels of candy to the point they were when the machines were installed (no higher than one-half full to a maximum of two-thirds full). Then simply go to the car, get the candy supplies from the trunk, and fill the machines.

And that is the successful completion of servicing your location. Now take your route cards, flip the one you just filled out up, over, and underneath the other cards, and there is your next location to service.

There are always exceptions to the actions you will have to take at the service locations, but for the most part, you have done everything required.

If your machines have coin mechanisms that are made of metal, I recommend that you bring into the store with you the can of spray lubricant every second or third month and spray the area on the inside of the coin mechanism that holds the coin into place when it is inserted into the mechanism. Just give this area a very quick spray. This will prevent the dirt on the coins from building up and jamming the mechanism. If your coin mechanisms are plastic and if you suspect a dirty coin track, you need to take off the back plate of the mechanism and clean the track that the coin runs through with your wet rag. Then be sure to dry it with the dry rag.

Also, if you have an iron-type stand that has been painted black, you need to spray the base of the stand with the spray lubricant to prevent it from rusting. This is especially important to be done monthly in restaurant locations where the floors are mopped quite often and water from the mop is constantly being put onto the base of the stand. Spraying the base of the stand with your WD-40 spray lubricant will prevent the rusting. If the base of your stand is a brown-painted metal, *lightly* spray some WD-40 onto it and wipe the spray down to cover all areas of the base with your dry rag. If your stands are the chrome-plated TV-stand type, just be aware of the bottom six inches of the legs of the stand. Wipe them down with your dry rag to remove dirt and dust. Use your wet rag, if needed, and follow up with the dry rag.

Basically, work to keep this business simple. Keep minimum tools, candy supplies, and other items down in number. You will find your time spent at each location to be less and less as you gain a better ability to work with your machines. You will begin to have a familiarity with the candy movement at each location. Some locations you will naturally fill up with more candy than other locations, just because you have realized that certain locations use more candy than others. These are the areas that you will come to know quite well. Be careful to never leave your machines at any location looking dirty or even dusty. Remember that you are working with a "high impulse" item and the success of the impulse is directly affected by the appearance of your machines, stands, and candy.

Just one last word: If you ever have to get inside the globe to handle any problem, this will most probably make your hand come in direct contact with the candy inside the globe. For health and sanitation reasons, never touch the candy with the hands. Under normal filling procedures, you do not have to touch any of the candy with your hand. You just open the bags of candy and pour the candy into the globe. For health and sanitation reasons, **NEVER** touch the candy with your hands. Have a few disposable powderless plastic gloves on hand. These gloves can be bought very inexpensively at a local supermarket or drug store. Buy the disposable type (preferably with no powder on them), and put a disposable glove on the hand that will be put into the machine to fix the problem. Do not let the candy touch your skin at all. It is important to remember that sanitation must remain your *number one concern* at all times.

Seven

Creating Routes and Route-Timing

I. CREATING ROUTES

Now that you have pounded the pavement and started the eternal cycle of getting the machines placed a little at a time, you are well on your way to starting a monthly income that can continue to increase with the amount of energy you put into this business. I can guarantee that you will be getting increasingly excited about the volume of coins (turned to dollars) that you will get **monthly** in direct proportion to the number of machines you get into locations. And the best part is that there virtually is no limit. Whether three machines or three thousand machines, the "PRINCIPLE OF DUPLICATION" is the same. Servicing machines, as well as keeping track with the proper paperwork, are all the same actions duplicated over and over again. Just duplicate your actions in these areas, and you can build a huge, bulk vending machine business.

Whether you eventually need more help to service all your machines in one month's time, your working space is too small, all of these problems are good ones. They show that your business is growing and promises a better future for you. There are solutions to any expansion problems you might experience. I will show you how to handle each problem in this and future chapters. I do not speak of the problems to discourage you, but rather to educate you, so you will be able to easily handle the small adversities that might possibly come your way. Hopefully, they won't, but if they do, you will know what to do. Let's begin with the first problem that you might run into.

You might find that you have machines scattered in different areas around town, and you are not quite sure of the route or path you should take in order to service them.

Find a good street map that has good detail of all the streets around the majority of the areas that your machines are located in. With the street map spread out on a table top, take a red pen and put a small "X" to mark every spot where one of your locations is. Now observe closely any areas that have several locations within them. This will be the start of your first two or three routes. A route is never completed. You will notice that routes continue to grow in size. You should group locations together in this style to form servicing routes. To easily identify the routes, name them after the name of the predominant street name that the majority of these locations are closest to. If these few locations are not located on any one particular street, adopt, perhaps, a town or city name. You will see the locations starting to fall into groups. These are **ROUTES**.

List each group of locations on a different piece of paper. List the locations in each route group in the best geographic order that you will use to service the locations. Now, arrange your route cards in the same order that is on the different sheets of paper. One sheet of paper equals one route. Likewise, each group of route cards equals one route. Bind each grouping of route cards with the two rings at the top of the cards, and there are your routes, all laid out and easy to follow.

List those locations that *do not* fall within the other route groupings separately. Have all of the corresponding route cards for these in one batch. Now, these "ODD" locations will be the first directions that you, or your locator, will go to fill in these odd areas with additional locations. Try to build these odd areas to be full routes as quickly as possible. These odd areas are the next areas of town that you will need to work to fill up with additional locations. This is how you build up your routes, and the direction that you will work in, to increase locations, in the direction that needs it the most. So consider that one isolated location is the start of a new route.

II. ROUTE-TIMING

Route-timing means timing each action so that you apportion and designate specific times of each month (or machine servicing cycle) to successfully rotate your work to include each location of each route in a smooth, regular pattern.

At this point, we begin to get into some slightly advanced areas of bulk vending machine usage. In order to get the most time and maximum exposure to all of your location routes, you need to plan which route you will work on which day of the month. You also need to be able to predict the approximate amount of time that each route should take you to complete. The accuracy of your "time to service" should remain somewhat flexible, not rigid. The reason you need flexibility in your predictions of needed time is that occasionally you will run across some repairs that are needed at a location, and you must take additional time, right then, to fix the machine. The repairs on the machines are not difficult, but they do take up time. So plan on a flexible basis the amount of time that it will take you to service each route.

Base your "time to service" on the amount of machines that you will total up within each of your routes. As you begin to service your machines regularly, you will have a good idea of how much time each machine takes. Then simply multiply this time by the number of machines that are at all locations within one route, and now you know how long it should take you to service this route. This amount of time will be needed for this route each month.

On a calendar designate the amount of hours or days that it will take you to service this route. Then those hours or days, each month, become **MAIN STREET ROUTE** time. Do the same thing for each of your other routes. This allows you to easily adapt the vending machine business into your life. It also promotes predictability within your business and into your "vending" time spent each day, week, month.

A) Plan each route in succession. For example, start from your most easterly located route and work toward your most westerly route. Or start on your most northern route and work your way to your most southern route. Get this kind of predictable control into your route servicing.

B) Get order and organization into it. This will save you countless amounts of travel time and gas expense. These actions will also give you the highest ratio of "time : money collected." Minimum time + maximum collections is the most ideal situation to work toward at all times.

Eight

Handling Situations

Occasionally, you will run into some situation that will require you to make some correction. Don't panic! You can handle it. Let's go over a few areas that you need to be aware of. This way if any of these things happen to you, you will know what to do.

You might experience low sales volume at a location. There are a few reasons that can cause low sales volume. Observation is the key here. If this should happen at a location, check on the following reasons:

Check to see if the business has a small candy area in which the store sells the same kind of candy that you have in the machines. If so, change what is in the globe.

Your machines might have to be moved more directly into the customer traffic area of the store. Perhaps closer, if possible, to the register.

The employees, or customers, might have grown tired of the candy selection that is in the machines. Try to change one or two of the selections. (Caution: be sure your candy portioners have been reset to vend the correct amount of candy pieces.)

Perhaps there just are not enough customers who walk into the store, i.e., the business does mostly telephone orders or the business's products are delivered to the customers.

You are experiencing candy spoilage. Candy spoilage is an occupational hazard in this business. The trick is to keep it to a minimum. There are several ways that you can do this, which follow.

Your machines were moved. Remain constantly aware of the location of your machines within the business. Don't forget about the greenhouse effect that your candy globes have upon the candy inside them. Sometimes someone will take the liberty of moving your machines from one spot to another because the management is rearranging the sales floor or sales merchandise. The candy might be in the direct sunlight or close to some machine or display case that emits too much heat for the candy. Then you will see, for example, the M&Ms starting to split open and the chocolate starting to turn to dust. Just ask if you can relocate your machines out of the direct sunlight or away from the source of heat. You can also change the candy to one that is more tolerant of the heat.

Perhaps the candy is starting to fade in color or starting to stick to each other. Again, too much heat. These problems are easy to solve.

"The candy is stale." As mentioned earlier, all candy usually has an expiration date marked on the box that it comes in. If your candy is growing stale, then it will have to be replaced. If the candy is liked by the employees and appears to be selling at a good pace, then decrease the amount of the candy that you allow to be put into the globe. This will allow for a fresher supply to constantly be available. Stale candy shows that the movement is usually slow and the amount in the globe is too high. You might possibly want to change this type of candy to another kind and monitor the new candy's sales rate.

Rather than throwing out otherwise good candy, try to make some phone calls around your neighborhood. Call stores like bakeries, ice-cream shops, or other food shops, and see if you can find a business that will buy your split or cracked M&Ms or discolored or cracked candies at a reduced price. I recommend a price that is one-third off the amount that you paid for the candy. The food shops might be able to use the candy in their baking doughnuts, toppings for ice cream sundaes, or any one of many other uses. This is an excellent way to get rid of cracked candy.

If the candy is really seriously beyond consumption, throw it away and list it as a loss for income tax purposes. Do not take any chances with stale candy. If it tastes bad, throw it out. Do not try to sell it.

Employees or customers complain that "only one or two pieces of candy come out for the quarter." This usually indicates that one or two pieces of candy are getting stuck between the springs or somewhere in the proportioner, and because of this, the proper amount of candy pieces is being blocked from dropping. Check out the springs for pieces of candy that are possibly stuck. The fruit jells like Mike & Ike or Hot Tamales are notorious for this. You will need to use one of your disposable plastic gloves and put your hand inside the globe to free the pieces of candy that are stuck in the springs. In the summer (or in direct sunlight places), the fruit jells get soft and have a tendency to smash very easily between the springs. If you have machines made predominantly of plastic, check around the outside of the rotating cylinders. The jells get smashed and squeeze around these cylinders. A clean wet rag will usually take the jells off easily. Total repair time needed—five minutes. Don't worry! You can handle it! This doesn't happen very often anyway.

"The machine will not take the coin." Or **"A coin is stuck!"** Usually when you hear this, something has been put into the coin mechanism that got stuck when someone tried to turn the handle. This can be fixed by removing the coin mechanism and freeing the item from the coin track. (Nine out of ten times, this problem is a bent coin or even a lead slug [a fake coin].) Be careful when you do this. Use one of the small screwdrivers if necessary. Do not pry too hard. It is always better to disassemble a part on the coin mechanism rather than try to pry the coin out. Prying can cause damage to the coin mechanism and then you might possibly need to order a new part. Again, this does not happen very often.

"There are ants in the machine." This is one of a vending machine owner's worst fears. Although inconvenient, this is really not so terrible. You will have to change the candy, clean the machine(s) with your wet and dry rags, refill the globes, and then take your roach and ant spray and spray around the base and pole of your stand to prevent this problem again. Then refill the globes. Upon future visits to this location, always put a little ant spray on the stand and pole. This will kill the ants before they reach the candy if they invade again. I spray a very small amount of ant spray on every location every month, just to be sure.

Well, here we have gone over 98 percent of the problems that you may possibly encounter with your vending machines. They are not very serious and usually can be fixed quickly. If these are the worst things that can happen, you can see that this business is a relatively worry-free business to own and operate. If you compare the above problems with the import/export business or with the twelve to fourteen hours a day and countless thousands of dollars you would have to put into a retail business, you can plainly see that the problems here are quite small.

Nine

Cut Out the Fat, Double the Dough!

Now that your *first phase* of machines have been placed into their locations and you are aware of the primary problems that bulk vending machine operators run across, you are officially on your way toward earning your own second **(monthly)** income.

This final phase will be what is known as "tighten up" or "cut out the fat." In this phase, you will be making improvements in various parts of your business based upon the results that you have been keeping track of from each machine in each location. These actions do not *have* to take place, but I strongly recommend they do.

Reevaluate types of candy in each location. Within three to six months, you will come to notice that certain candies or novelties seem to sell exceptionally well at certain locations. It will become equally apparent that other candies or novelties just do not seem to sell nearly fast enough at certain locations. It takes time and observation to become aware of these movement patterns. At the point that you have observed one certain candy has nearly half the coins in the coin box that the other coin boxes have at the same location, and this has continued for more than two months, then it is time to try another candy in place of the slow-moving candy. If this is done regularly, you will be finding more and more coins in those coin boxes.

Reevaluate how many machines should stay in each location. As an unofficial requirement, from my experience and the experience of many other bulk vending machine owners I have talked with, I would estimate that the **minimum** acceptable sales requirement for each machine in each location should be **approximately $7 to $10 per month**. I use this as a way to judge how good or bad each machine is doing. If you find all the machines in one location are below this amount and you have already tried changing candies, but sales stay about the same, then I recommend that you decrease the number of machines at this location. Do not remove the machines from the location. Each location is valuable. Just cut back on the number of machines in that location. Then you will have more machines to be placed at another location, without any more investment, and sales will increase because of this move.

This action alone will do more than one would think toward a large increase in the total sales within your business, without the extra investment for machines.

Reevaluate number of candy pieces being sold per coin. For one of several possible reasons, the number of candy pieces being vended per coin could change. Once every six months, give a quick check on the actual amount of candy being vended for each coin.

The amount being vended could change because:

The screws on the proportioner may have become loose.

You may have set the portioner dishes to a higher setting than recommended by the manufacturer. There is a tendency to do this with the first batch of machines that we put into our first few locations.

The candy company may have decreased the size of the candy.

Relocation services should *not* be overlooked. Look for new locations *within* your route areas six months or one year later. Some of your locations may go out of business and a new business will be in the same store front. Rental leases expire, businesses change their locations, and business owners sell their businesses to new owners. There are many reasons to recheck the same areas you have previously canvassed for locations. You will be pleasantly surprised with the results.

Don't be afraid to "cut out the fat" from your business operation. You will save money on candy, increase utilization, increase your sales, and enlarge the size of your business. All these things can usually be done with no additional investment. These things can be done while you are doing your location servicing routine and are the very actions that all businesses work the hardest at achieving. They strive to "close the gap," "tighten up," "cut out the fat." You too will be happy you did.

Ten

How Many?—How Big?

By following the information in the last nine chapters, you are now beginning to get a good feel for this business. At the six-month to one-year time mark, you will truly be able to start seeing some good income if you have worked at this business consistently. Now just imagine what you can do full-time in this business. As 10 or 15 locations grow into 50, 75, or even 100 locations (275 to 300 machines), you will start to see approximately $1,000 to $1,500 or more **per month**. As you continue to duplicate your efforts and continue to put out more and more machines in additional locations, you will see where major income can be achieved. The funny thing about this business is that the income possibilities are very deceiving and are not promoted at all within the public media.

As 100 locations turn into 400 to 500 locations over the next months or year, you will definitely feel the requirement for some larger working space in order to physically accommodate your business volume. This can be done easily within one mini warehouse approximately fifteen feet wide and thirty feet deep.

Plan for space enough to accommodate the following needed resources:

1. Machine purchases, assembly, and storage until they are located.
2. Storage space for your direct shipments of candy and novelties from different manufacturers.
3. A machine repair area, complete with a small inventory of spare parts.
4. A small office space where all the paperwork is handled, phones are answered, files are kept, purchase invoices and shipment receiving is handled, and payroll is done.
5. The vehicles needed to properly service all your locations in a timely manner.
6. Full-time locators and honest, reliable personnel to service your large volume of hundreds of vending routes.

I touch upon these areas only to try to give you the big picture. When you see how many coins are coming out of your machines, you will truly understand what can be done on a larger scale. Several people who I know have been in the bulk vending business for ten years or more and are now into real estate investing and development. What does that tell you? It means that there is plenty of money to be made in this business. The money is there for the taking if you want it. This can be developed into a family business where everyone is involved. If you are retired, this is a good way to utilize your time in a cash business.

Another area to consider within the vending business is the great amount of diversification that a person can get into. How about all those soda vending machines, snack machines (bar candy and chips), food machines with microwave ovens, cigarette and postage stamp machines? You could take some profits and start buying larger machines that vend items up to two dollars each. These machines have a dollar-bill acceptor on them. They will provide larger amounts of profit. The list just seems to be as endless as your imagination. It is all there, waiting for you. Nearly anyone can learn to fix the larger machines, just as you have learned to fix the bulk machines. The only thing you really need to do is to **GET STARTED**.

Eleven

Conclusion

The very first day that I came home from servicing a couple of my routes and I counted up the take, I was very excited to find that I had collected $480 that day, in only six hours. I was in no hurry to service my machines. I had already placed about 65 machines into store's at that point. Those machines had created about 21 or 22 different locations. The majority of my locations have 3 machines in each location. All of my machines are twenty-five-cent machines. What a thrill that was for me. I know what a thrill awaits you.

I just want to leave you with these final words. The things we work the hardest at having are the things worth having the most. It is important to run your business ethically and honestly. We expect these same things from our bosses, friends, family, and we teach these things to our children. Be sure you check with your appropriate state and local city agencies for proper licensing requirements. This business will be as good to you as your ethics are to it. State and city licensing requirements vary from state to state, town to town, city to city. Build your business so that it can endure the test of time and remain forever strong and profitable. *Build your business with a large amount of ethics.*

YOU CAN DO IT!
GOOD LUCK!!!!